Lucas Miguel Micaloski

Biotechnological applications, stem cell genetics

Lucas Miguel Micaloski

Biotechnological applications, stem cell genetics

ScienciaScripts

Imprint

Any brand names and product names mentioned in this book are subject to trademark, brand or patent protection and are trademarks or registered trademarks of their respective holders. The use of brand names, product names, common names, trade names, product descriptions etc. even without a particular marking in this work is in no way to be construed to mean that such names may be regarded as unrestricted in respect of trademark and brand protection legislation and could thus be used by anyone.

Cover image: www.ingimage.com

This book is a translation from the original published under ISBN 978-613-9-72671-4.

Publisher:
Sciencia Scripts
is a trademark of
Dodo Books Indian Ocean Ltd. and OmniScriptum S.R.L publishing group

120 High Road, East Finchley, London, N2 9ED, United Kingdom
Str. Armeneasca 28/1, office 1, Chisinau MD-2012, Republic of Moldova, Europe
Printed at: see last page
ISBN: 978-620-7-90294-1

Copyright © Lucas Miguel Micaloski
Copyright © 2024 Dodo Books Indian Ocean Ltd. and OmniScriptum S.R.L publishing group

SUMMARY

The fusion of the sperm and egg gives rise to the first stem cell, called totipotent. It has the potential to differentiate into all embryonic and extra-embryonic tissues. The different types of tissues occur through stimuli from the stem cells themselves, or from already differentiated cells, in which the conditioning of the environment leads to the interaction of surface proteins, promoting the activation and deactivation of genes, altering morphology and behaviour. After the totipotent cells differentiate into embryonic and extra-embryonic tissues, they lose part of their potential, differentiating into pluripotent stem cells, which can only give rise to embryonic tissues and stem cells of lesser potential.After the birth of the individual, the pluripotent cells present themselves in a more mature form, with their potential reduced, being called multipotent stem cells. Because stem cells have this potential for differentiation, they can be used in tissue engineering, regenerative treatments, some cancers, research aimed at obtaining greater cellular dominance, with the aim of curing degenerative diseases.There is still a need for a broader and more satisfactory understanding, where DNA activities are being studied that cause gene regulation, which is essential for cell differentiation, epigenetics being one of the forms of regulation; obtaining greater control of stem cells is evolving through studies and new applications such as CRIS/PR, hormonal studies, microRNAs, gene mapping, methylation and acetylation of genes, among others. It is believed that these cells will be widely used in biotechnological processes in the coming decades.

Keywords: Stem Cells.   Embryo. Biotechnology.
CRIS/PR. microRNAs. Hormonal studies.

SUMMARY

CHAPTER 1	3
CHAPTER 2	5
CHAPTER 3	6
CHAPTER 4	7
CHAPTER 5	14
CHAPTER 6	18
CHAPTER 7	27
CHAPTER 8	32
CHAPTER 9	33
CHAPTER 10	37
CHAPTER 11	40

CHAPTER 1

INTRODUCTION

This review aims to address issues in biotechnology regarding the characteristics, possible risks, advances and challenges in relation to stem cells, mentioning their behaviour and their indispensable role in the origin of an organism and after its origin (VILLAMARIN et al,2009; BYDLOWSKI et al,2009).

There are various applications that will make it possible to modernise medicine in the future by providing solutions for organ transplants produced in laboratories, degenerative cures for various diseases that are currently unsolved (DOMINGUES et al, 2016; TORRÃO et al).

As well as mentioning the possible advances, various problems will also be addressed, including which is the best source of stem cells, and what potential each lineage has for various applications, taking into account the best result and the implications for the law due to the controversies surrounding their manipulation (ABDELHAY et al, 2009).

The genetic characteristics of these cells were analysed, addressing the need for control and protection through the study of genes and the telomerase enzyme (GOMEZ et al, 2014; SOUZA et al, 2010).

The meaning of the name stem cell means that various types of branches originate from the stem, which would be differentiated cells or specialised in a certain function. Stem cells can be compared to the "joker" card, where they could replace the function of any card in various games. In the same way, depending on their potential, stem cells could take on the morphology of any cell in a given organism (BYDLOWSKI et al, 2009;

KABAYASHI et al, 2015).

This induction of stem cell differentiation occurs according to the stimulus it receives, leading to the activation or deactivation of genes and thus causing changes in cell morphology. The changes occur naturally in an organism or in an in vitro culture (ABDELHAY et al, 2009).

There are major challenges in stem cell research, including a huge amount of information that has yet to be discovered: the best type of stem cell to use in research and treatment, ways of inducing them to differentiate into tissues, eliminating possible risks and possible application in neurological and cardiovascular treatments, tissue engineering and other applications (BARBANTI et al, 2005; RODRIGUES et al, 2012; BARETO et al, 2015).

In addition to the challenges relating to the domain, there are also problems in releasing research because of bioethical polemics and religious thinking regarding the destruction of embryos to obtain embryonic stem cells, which has been opposed by religious organisations; there is also a lack of information, because the same embryo destroyed for scientific purposes would already be destroyed for disposal (MIRANDA et al, 2013).

CHAPTER 2

BACKGROUND

Stem cells have a major influence on animal development up to the adult stage. More research into the fundamental principles of stem cells is needed, as well as a list of their risks and misuse, in order to scientifically publicise their potential (KABAYASHI et al, 2015; PINHO et al, 2009).

Showing possible clinical therapeutic and tissue engineering applications, which will bring about a possible medical revolution in the future. In the next few years, it is possible that we will achieve a cure for diseases such as Alzheimer's, Parkinson's, diabetes and various other degenerative diseases. In addition to these therapeutic treatments, stem cells have the potential to produce organs in the laboratory, which will greatly contribute to transplants made with the patient's own genetic material, reducing rejections (DOMINGUES et al, 2016; ALDELHAY et al, 2009; RODRIGUES et al, 2012).

CHAPTER 3

GENERAL OBJECTIVE

Unifying various pieces of information on biotechnological applications relating to stem cells in a single work, mediating a certain cellular dominance in some applications and a lack of dominance in others.

3.1   SPECIFIC OBJECTIVES

- To analyse bibliographies on stem cells, using scientific electronic portals to search for publications.
- Identify advances in the application, procurement and control of stem cells and their possible clinical uses.
- Describe the interaction of the cellular microenvironment in altering the behaviour and morphology of stem cells, which are responsible for cell induction or non-induction.
- Mention ways of activating and deactivating genes and their results.
- Expose the relationship between genetics and epigenetics resulting in new cell behaviours from a stem cell to a somatic cell.

CHAPTER 4

ORIGIN AND DEVELOPMENT OF STEM CELLS

Its origin occurs through the fusion between the sperm and the fertilised egg, this is the first cell of this possible individual, being called a totipotent stem cell. It replicates several times, resulting in a cluster of totipotent cells called a morula, with the potential for each cell in this cluster to give rise to a new individual (VILLAMARIN et al, 2009; BYDLOWSKI et al, 2009).

These cells have no specialisation or differentiation, but with a specific induction they have the ability to specialise into any cell lineage, giving rise to the embryonic leaflets mesoderm, endoderm, ectoderm, germ cells, trophoblast, among them the blastocyst, which gives rise to pluripotent stem cells, reducing their differentiation potential. They can only differentiate into embryonic tissues (BYDLOWSKI et al, 2009; JUNIOR et al, 2009).

After the foetus is formed, these cells become more mature and multipotent, subdividing into haematopoietic and mesenchymal. They have the potential to differentiate according to their location, with the function of building and regenerating tissues (BYDLOWSKI et al, 2009).

Both stem cells have the power to renew themselves, and one stem cell can give rise to two stem cells of the same power (symmetrical division) or of different power or to specialised cells forming tissues, or one of the daughter cells may have the same potential as the mother cell and the other may not (asymmetrical division) (BYDLOWSKI et al,2009; JUNIOR et al,2009).

The behaviour of stem cells depends on their microenvironment, where they occur through secretions from the stem cell itself of the same potency, different potencies or even from already differentiated cells, forming interactions in the form of stimuli that lead the cell only to replicate itself or to induce itself in other types of cells, through signalling molecules called cytokines (groups of molecules made up of protein and other molecules linked to it, whose function is signalling).

Expressed by various cells, they work as a communicator by binding to surface receptors, which can promote a certain command that will make the stem cell replicate or activate or deactivate genes leading the stem cell to specialise in a particular tissue (PINHO et al, 2009; BYDLOWSKI et al,2009; JUNIOR et al,2009).Since the origin of the embryo, there has been a series of differentiation of stem cells into other cells, but in order for this alteration of the cell to occur, something is needed to induce it, which can be done in various ways; In pregnancy, this process is known as embryogenesis, in the way that the cell membrane stimulates the binding of mediators to their receptors, the cytoplasm stimulated by the membrane transmits to the nucleus activating transcription factors, where genes will be expressed for the production of proteins that will determine the behaviour of the cell (PINHO et al, 2009).

4.1 HAEMATOPOIETIC STEM CELLS

Progenitor of multipotent cells, they originate in the embryo, producing various tissues and being maintained after the birth of the foetus where they act in the regeneration of some tissues and in the production of short-lived cells. Haematopoietic stem cells are in low concentration in mammalian organisms after birth, only found in the bone marrow at approximately 0.01% to 0.05% and in the peripheral blood at less than 0.001% (ABDELHAY et al, 2009).

Three types are known: LT-CTH, which are capable of long-term self-renewal and are indefinitely responsible for producing cells for the entire haematopoietic system, CT-CTH, which are capable of short-term self-renewal, and PM, which are incapable of renewing themselves (JUNIOR et al, 2009). They have limited power to differentiate into blood tissue and immune system cells, with haematopoietic stem cells differentiating into common myeloid cells (erythrocytes, platelets, dendritic cells, macrophages, neutrophils, eosinophils, basophils) and lymphoid cells (B and T lymphocytes, dendritic cells and NK cells) (PAYAO et al,2009; VOGORITO et al,2009; VISENTAINER et al,2008).

In haematopoiesis, regulation occurs through the microenvironment, by interaction between other types of cells that secrete substances such as cytokines. These cell communicators have an affinity for haematopoietic surface receptors, promoting regulation of various functions such as quiescence, self-renewal, differentiation, apoptosis and mobility. When exposed to IL-3, IL-6 stimuli *(interleukins),* SCF, TPO, FLT-3 (signalling protein molecules), the haematopoietic cells are stimulated to different functions (ABDELHAY et al, 2009).

Many genes influence the kinetics of haematopoietic stem cells, the most important of which are genes that encode transcription factors (hoxb4), genes that encode signalling molecules (haematopoietic cytokines, WNT family proteins, *Notch family* receptors*),* and genes that regulate the cell cycle.

Hoxb4 has the ability to increase the self-renewal of haematopoietic cells, while the *WNT* signaler (a molecule that *binds to* Frizzled receptors and stimulates cell proliferation), when absent, undergoes *proteosomal* degradation of intracytoplasmic *b-catenin,* after the WNT ligands signal, *b-catenin* degradation is inhibited (protein-beta-catenin complex) and its accumulation translocates to the nucleus with the transcription factor TCF (family of DNA-binding proteins), regulating various genes. Several *WNT* and frizzled proteins are expressed indicating paracrine and autocrine effects of the ligands such as *WNT5A, WNT2B, WNT10B. On the* other hand, *P16/ARF* (cyclin-dependent kinase 2A inhibitor, multiple tumour suppressor 1) inhibits haematopoietic proliferation by inducing *apoptosis,* which is reversed by the expression of Bmi1 which is a repressor of *P16/ARF* (ABDELHAY et al, 2009).

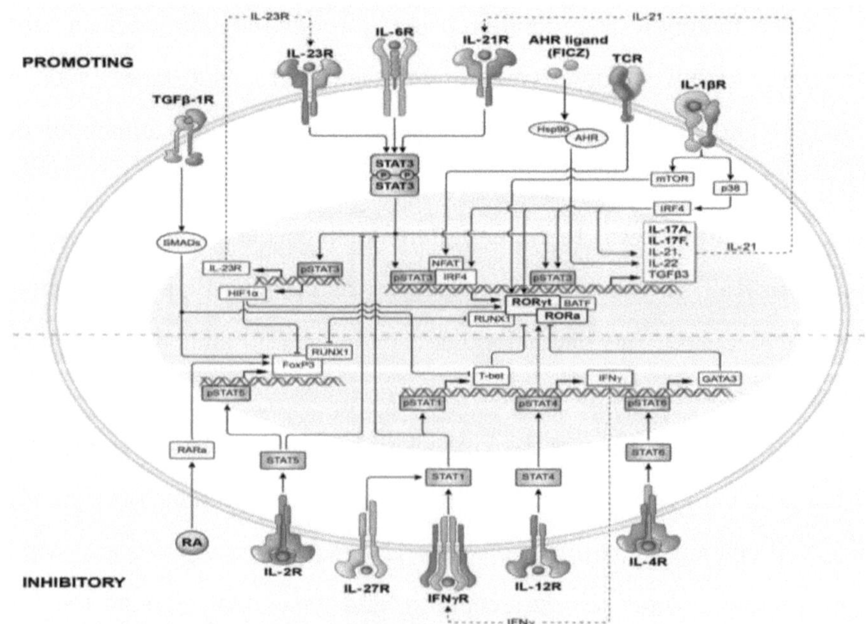

Figure 1 - Kinetics of Haematopoietic Stem Cells
Source: ABDELHAY et al, 2009.

4.2 MESENCHYMAL STEM CELLS

They are multipotent progenitors, are found in most adult tissues and have the capacity for self-renewal and specialisation in all cells of mesodermal origin between cartilage, bone and adipocytes, providing support in different tissues such as bone marrow and adipose tissue. They are cells that have a great affinity for inflammation and can even migrate to damaged tissues, secreting cytokines with antibiotic and anti-inflammatory properties and growth factors through bioactive molecules (BYDLOWSKI et al, 2009).

They have regenerative action in human clinical use for the treatment of skin burns, ulcerations of the lower limbs and ophthalmic lesions. Because stem cells have immunosuppressive capabilities, self-renewal and differentiation into cells of various tissues, they have a wide variety of uses in therapies in which exogenous cells are implanted into damaged tissues and systems, treating diseases such as diabetes, ischaemia, neurological diseases and many others (KABAYASKI, 2015; BYDLOWSKI, 2009; COLPO et al, 2015).

Mesenchymal stem cells can release or receive secretions in response to pro-inflammatory stimuli, hypoxic stimuli, exposure to apoptosis factors, directly or indirectly in injured tissues, acting in the formation of new blood vessels, nervous systems, regulating the environment, promoting protection and repair processes. Among the secretions are cytosines, antioxidants, extracellular matrix proteins and hormones. It has been observed that hormones secreted by mesenchymal stem cells act by conditioning the environment and protecting neurons, even at a certain distance, with a neuroprotective effect through neurotrophic factors (COLPO et al, 2015; MONTEIRO et al, 2010).

Mesenchymal cells can specialise into chondrocytes through TGF factors, the most potent of which is IGF-1, BFGF, EGF, PDGF and Wnt (signalling molecules). In the production of cartilage, the production of cell matrix protein and collagen II is stimulated through the intracellular signalling pathways MAPK and Smads, which activate transcription factors SOX9, SOX5, SOX6 (gene-binding molecules). The induction of osteogenic or neural cells occurs with the stimulation of BFGF and endothelial Wnt is formed through the stimulation of VEGF (vascular *endothelial* growth factor) (PINHO et al, 2009).

In vitro culture of mesenchymal stem cells differentiating into osteoblastic cells, the following results were obtained: using DEX (dexamethasone) and ascorbic acid, the mesenchymal cells begin to develop obtaining an osteoblastic morphology, the production of alkaline phosphotase and expressions of bone matrix protein mRNA and accumulation of mineralised hydroxyapatite were observed, after 16 of culture in these conditions, an accumulation of calcified matrix appears on the surface of the plate, confirmed by von kossa staining. This induction occurs through DEX and a cofactor, ascorbic acid, which hydrolyses proline and lysine residues in collagen, increasing the production of non-collagenous bone protein matrices (BYDLOWSKI et al, 2009; PAYAO et al, 2009). Application in neurodegenerative diseases has achieved effective results, where in addition to regulating inflammation, it causes benefits beyond the objective, such as neuronal growth, reduction of free radicals and apoptosis (LIMA et al, 2012; SILLA et al, 2010).

Studies show that mesenchymal stem cells have the ability to migrate to injured sites through mechanisms that involve the expression of specific receptors that facilitate their arrival in affected tissues. The way in which mesenchymal stem cells reach injured tissue is called *"Oorning"*, one form of application is intravenous, and the efficiency of the treatment will depend on the greatest possible number of mesenchymal stem cells reaching the target site (LIMA et al, 2012).

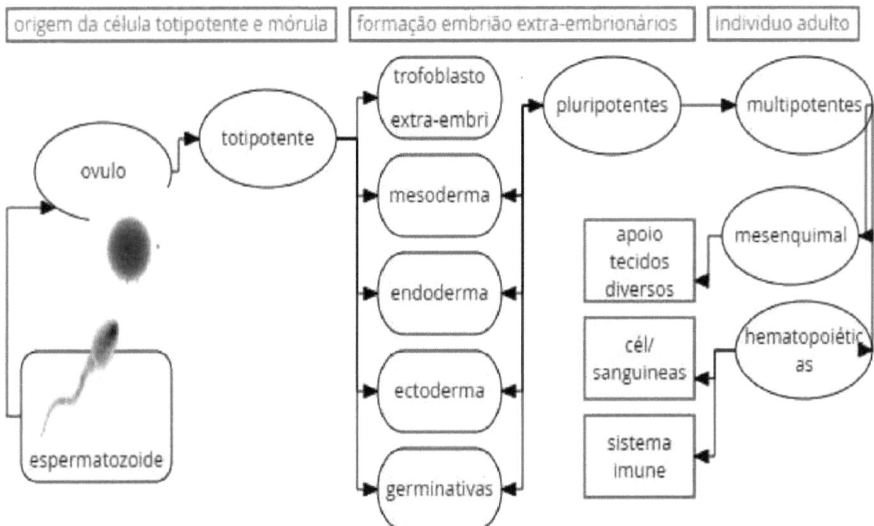

Figure 2 - Stem Cell Kinetics
Source: MICALOSKI, Lucas Miguel

CHAPTER 5

CULTURES AND GENETIC CHARACTERISTICS OF STEM CELLS.

The cultivation of stem cells requires criteria of quality, reliability and traceability, so that there is no unwanted differentiation or loss of function, such as inducing cancer. Embryonic stem cells are rarer, while mesenchymal stem cells are easier to access, but because they are adult stem cells they are difficult to culture and sometimes the enzyme responsible for immortalisation, telomerase, can be lost.

In vitro, two aspects can occur: a reduction in the lifespan of stem cells, which enter senescence and then apoptosis, and the crisis phase and neoplastic transformation. Senescence is associated with the shortening of telomeres and after this the cell cycle is interrupted and apoptosis can later occur. In the neoplastic phase, structural changes may occur in the chromosomes, such as deletions, translocations and inversions, leading to abnormal expression of various genes, loss of cell speciality and the development of possible cancer. In order to monitor these cells, it is necessary to apply G-banding, SKY, through culture and freezing from the first to the twentieth passages. G-banding is used to detect numerical and structural changes in the chromosomes obtained during culture. The SKY technique has the function of characterising gains or losses of chromosomal material with a high resolution of around 1MB (GOMEZ et al, 2014; ALVES et al, 2017; PERINI et al, 2008).

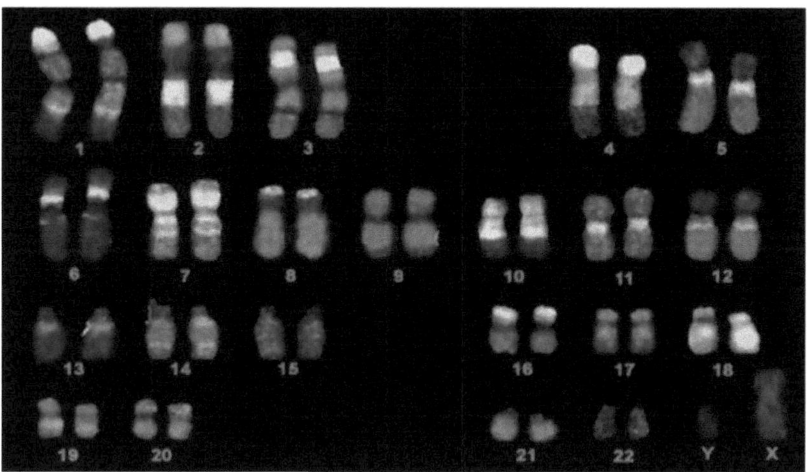

Figure 3 - G-banding
Source: text&image.

5.1 TELOMERASE ENZYME

At the linear end of vertebrate DNA are telomeres, which are several repeats of nitrogenous bases whose main function is to protect the chromosome, preventing it from being damaged. With each mitosis, the telomere gets shorter due to the loss of bases. In a somatic cell, after 54 mitoses, the telomere ends, causing the cell to enter senescence and perhaps apoptosis, leading to tissue degeneration and functioning as a biological clock (MIRANDA et al, 2013; PERINI et al, 2008).

Some cells, such as stem cells, germ cells and tumour cells, have telomere protection in which an enzyme called telomerase recovers lost telomere sequences. In stem cells, the action of these telomerase enzymes is stronger in totipotent cells and gradually weakens until they become multipotent cells, especially haematopoietic cells which are very susceptible to telomere loss due to low telomerase action (GOMEZ et al, 2014; PERINI et al, 2008; SANMARTÍN et al, 2014).

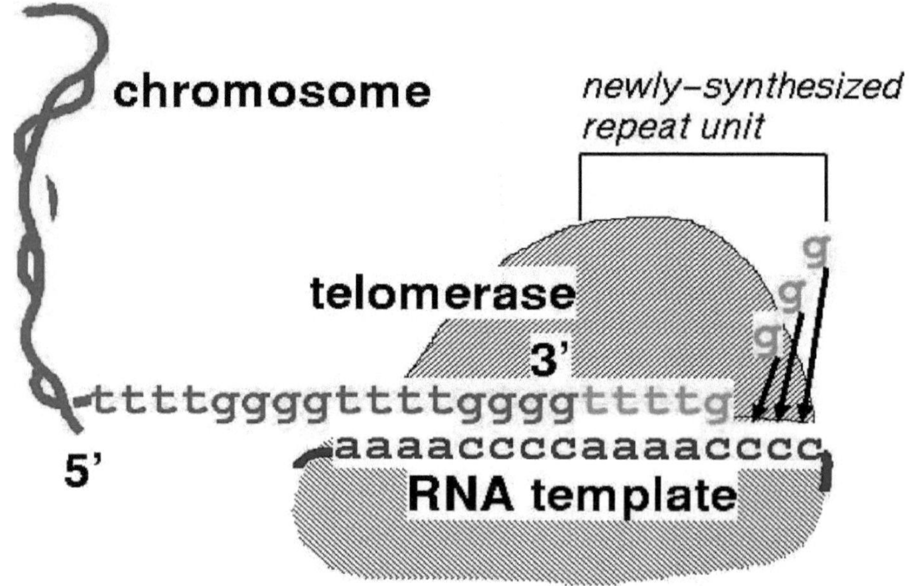

Figure 4 - Telomerase enzyme
Source: BOX, Jessica A; BUNCH, Jeremy; et al.

5.2 CANCER STEM CELL

Evidence points to the existence of cancer stem cells, and in several characteristics the two are similar, such as the ability to self-renew, the possibility of differentiation, high expression of telomerase, evasion of apoptosis, increased transport of substances across the membrane and the ability to migrate. There are three hypotheses for the origin of this cancer stem cell, in which multipotent stem cells mutate into cancer stem cells, another hypothesis would be that an already differentiated cell undergoes a reverse induction giving them pluripotency, and the last hypothesis would be a rare fusion between multipotent cells and other cells (KABAYASHI ET AL, 2015).

Unlike multipotent stem cells, these cancer stem cells have unlimited dividing power and are resistant to chemotherapy and radiotherapy, so even after surgery to remove a presumed tumour, there are likely to be metastases and more tumours elsewhere (KABAYASHI et al, 2015; BUNCH et al, 2008; MAGALHAENS et al, 2014; WALTERO et al, 2014).

CHAPTER 6

CELLULAR INFLUENCERS AND GENE ACTIVATION AND SILENCING

The formation of tissues and cell regulation is dependent on the activation and deactivation of certain genes; where morphological changes occur in the cell and the production of new cell products without altering the genetic code. This differentiation occurs in ways in which the phenotype regulates the genotype (without altering the code), the genotype regulates the genotype (without altering the code) and as a result the genotype responds to the phenotype by forming tissue assemblies or certain metabolic assistance. These events, which result in changes in cell morphology through the activation and deactivation of genes, are today known and studied as epigenetics, which occurs through the acetylation and methylation of genes, microsRNA, and the formation of histones.

6.1 EPIGENETICS

Epigenetics is the influence of various reversible and irreversible changes in the expression of a gene, but without altering the gene code. This epigenetic influence occurs through environmental and behavioural stimuli, according to Benitez et al (2015) diet and correct exercise promote the production of hormones that directly or indirectly have different effects on children's growth, lifestyle, physical exposure to sunlight, radiation and chemicals.

According to Sandoval et al (2018), alcohol consumption results in epigenetic changes in DNA methylation, tobacco and can also occur internally through hormones. Non-coding RNA (microRNA), gene methylation, alteration of histones, can then conclude that epigenetics is the interaction of the phenotype in the genotype and the genotype in the genotype.This mechanism occurs since the formation of the zygote and is very important in cell differentiation and tissue formation through the activation and deactivation of genes, such as the X chromosome that undergoes several deactivations; it is also important in the ultra-uterine adaptation of the organism, acting as a form of cellular plasticity. These epigenetic events can occur in somatic cells, which are reversible, or in germ cells, which can be passed on to future offspring (ORNELLAS et al, 2017).

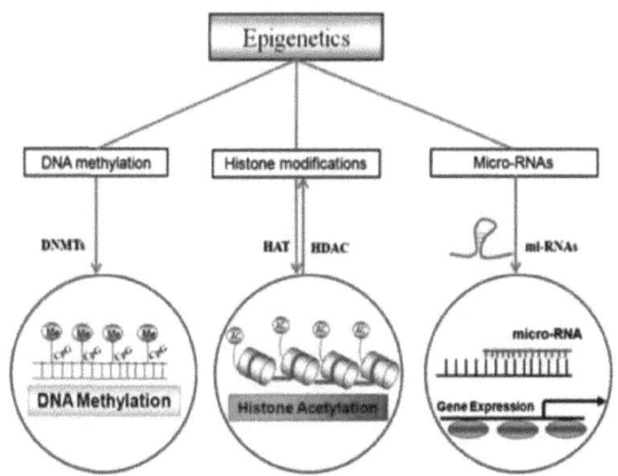

Figure 5 - Epigenetic mechanism
Source: ELORRIAGA, Tomas.

There are several epigenetic mechanisms, the most studied of which are: being directly in the DNA, promoting the inaccessibility of genes through deacetylation and methylation of specific DNA sites; tangling of DNA in histones and their proteins that form it, giving rise to condensed chromatids; the reverse also occurs, with many substances promoting the reverse above, with acetylation and deacetylation; remodelling of the chromatid. After transcription, epigenetics also occurs through microRNAs that act on mRNAs, preventing their translation (AFANADOR et al, 2018).

6.2 HORMONES AND GENE REGULATION

Hormones are divided into five classes: vitamin derivatives: retinoids and vitamin D; amino acids: dopamine, catecholamines and thyroid hormones; neuropeptides:GnRH, GHRH, TRH, somatostatin (SMS), vasopressin; proteins: insulin, LH, FSH, TSH, Hcg, PTH, ACTH, growth factor peptides; steroids: cortisol, aldosterone, estrogen, androgens synthesised from cholesterol, progesterone. Hormones derived from amino acids, peptides and proteins interact with membrane receptors. Thyroid hormones, steroids, vitamin D and retinoids are piposoluble and interact with intracellular nuclear receptors (FONTENELE et al, 2010).

In terms of hormone receptors, there are four membrane receptors: trans-membrane G-protein-linked, tyrosine kinase, cytokine kinase, serine kinase; and two nuclear receptors: 1 and 2. Hormones have the functions of being regulators where they dictate the rhythm of the entire metabolism and messengers where they can trigger a change in cellular behaviour by transcribing new proteins and blocking others, and can even alter the morphology of the cell, making it specialised in some tissue (FONTENELE et al, 2010). An example of differentiation through hormones is the production of T-cells in the thymus organ through hormones, including the hormone thymulin, which is related to the differentiation of haematopoietic cells into T-cells, one of those responsible for immunity in the body (FAUSTO; ET AL2004). They are further specified in the tables on the next page, and are related by group, receptor, class, hormone and function:

Table 1 - Regulation and change of cellular behaviour through some hormones - intracellular

Group	Receiver	Class	Gland	Hormone	Function	Target cells
Nuclear	1	Steroids	Adrenal cortex	Glucocorticoids	Immunosuppressant	Liver and muscles.
			Supra-renal	Mineralocorticoid	Stimulates transcription, affects the expression of specific genes, electrolyte balance.	Kidney,
			Ovaries	Estrogen	Development of the female body and its characteristics.	Breasts, waist, uterus...
			Corpus luteum	Progesterone	Pregnancy hormone acts on the entire female physique, preparing it for pregnancy.	Endometrial cells.
	2	Amino acid derivatives	Tirióde and paratirióde	Thyroid	T3 and T4 Regulate metabolism, development, maturation and CA+ regulation in the blood, preventing bone decalcification. Pth increases the concentration of calcium in the blood, demineralising bones.	All the body's organs.
		Vitamin derivatives	Skin	Vitamin D	Maintenance of calcium levels in the blood and bones, metabolic, muscular, cardiac and neurological functions. Cell growth, differentiation and apoptosis.	Bones, kidneys,
			Vitamin A derivative.	AcidRetinoic	Growth and development activating mitosis and cell renewal, cell differentiation,	Collagen, elastin
					embryogenesis, bone development.	

Box 2 - Regulation and change of cellular behaviour through some hormones - extracellular

Group	Class	gland	Hormone	Function	Target cells
Membrane	Proteins	Hypophysis	IL	Regulates androgen sex hormones.	
			FSH	Development, growth and maturation of the gonads	
		Adenohypophysis	TSH	Stimulates the thyroid	Thyroid
		Trophoblast syncytium	HCG	Maintaining the corpus luteum in the ovary	Trophoblast

6.3 MICRORNA TRANSLATION INHIBITORS

MicroRNAs are small fragments of RNA, approximately 20 nucleotides long; they do not code for proteins, but rather inhibit the translation of certain proteins, and their coding is an anti-sense mechanism of the mRNA, enabling it to fit together. Their function includes cell differentiation, apoptosis, proliferation, development, embryonic tissue formation and cell cycle control (CAVAGNARI,2012; GAMBOA et al, 2014).

The transcription of microRNA occurs with the formation of primary RNA, which is initially originated by the enzyme RNA polymerase I or II and then finalised by a complex microprocessor made up of drosha proteins with their associated proteins (DGCR/pasha), processing primary RNA with approximately 70 nucleotides, after which this primary RNA is exported (aided by exportin-5) to the cytoplasm where it matures to form double-stranded microRNA with approximately 21 nucleotides produced by the enzyme RNAse III Dicer and its associated TRBP protein.

This microRNA has its code compatible with a certain messenger RNA where it can bind and suppress that particular gene expression, preventing translation from taking place, and at some point the mRNA is degraded (GAMBOA et al, 2014).

It is estimated that microRNAs account for 3% of the human genome and are responsible for regulating 60% of genes, with just one microRNA being able to regulate approximately 200 different transcripts, each acting in different cellular pathways, and mRNA being regulated by various types of microRNA.

Some microRNAs also act as methylation guides for certain genes (they are also gene regulators) and can also have a positive or negative effect depending on cell differentiation (GINER et al, 2013).

It has certain oligonucleotide enzymes called anti-microRNA, which function as targets and are microRNA silencers. They are synthetic anti-sense and competitively inhibit the interaction of microRNA and target mRNA (GINER et al,2013).

6.4 METHYLATION AND GENE SEALING

It occurs in eukaryotes; it is a chemical reaction characterised by the addition of a methyl group in a specific position, being C5 on the cytosine ring, which is catalysed by the enzyme DNA methyltransferases, resulting in the formation of 5-methylcytosine. This methylation is essential in cell differentiation and plays a major role in the formation of the embryo, promoting the switching off of specific genes such as the X chromosome, one of which is inactivated in the case of females, and is essential in gene regulation, genomic presses and chromatin modifications.

Methylation is frequent in cytosine-phosphate-guanine (CpG)-containing regions, 55% of which are in the promoter regions of approximately 40% of mammalian genes; factors will cause methylation to occur or not, with hypermethylation in promoter regions switching off gene expression and hypomethylation activating the promoter region and consequently possible gene expression.

Other proteins will bind to this modified 5-methylcytidine region, such as adipocyte protein-2 (AP-2), Cmyc/murine max homologue (Cmyc/myn), cyclic adenosine monophosphate response element binding proteins (CREB), E2 factors (E2F), nuclear factor-KB (NF-KB). These binding sites may be connected by other proteins such as methyl-CpG 2 (MeCP-2), methyl-CpG binding domain proteins (MBD) 1, MBD2, MBD3, and MBD4, binding to methylated cytosines stimulating chromatin condensation resulting in gene inactivation (ORNELLAS et al, 2017).

The enzymes that participate in the addition of methyl groups to cytosine molecules are from the DNA methyltransferases (DNMTs) family, being DNMT-1, DNMT3B, DNMT3A and its isoforms, and DNMT3L; where DNMT1 is responsible for maintaining methylation patterns during mitosis (ORNELLAS et al, 2017).

But how is this methylation passed on to new generations of cells and, in the case of germ cells, to their descendants? Cytosines are in sequences with a methylation site being 5'GpC3' and its complementary 5'CpG3'. Both chains can be methylated while maintaining methylation in each replication in an original strand, at which point the DNA methyltransferase molecules (mentioned in the previous paragraph) interfere by recognising the 5'methyl-CpG3' sequence and methylating the opposite complementary unmethylated chain, making it possible for new future daughter cells to be methylated (CAVAGNARI, 2012).

It is worth mentioning the importance of studying gene methylation, which enables therapeutic treatments that can activate or deactivate the transcription of genes that will bring different results. One example is the development of the drug azacitidine, which has been evaluated for the treatment of myelodysplastic syndromes (CAVAGNARI, 2012).

6.5 HISTONES AND GENE ACCESSIBILITY

Occurring in eukaryotes, they are extremely important in gene regulation, promoting accessibility and non-accessibility of genes, resulting in gene expression or non-expression depending on chromatin remodelling. Histones are proteins that work as if they were DNA strand spools, after part of the DNA is wrapped around the histones, nucleosome proteins bind to the histones, condensing the chromatid (ORNELLAS et al, 2017).

Histones have troughs where events such as acetylation, methylation, phosphorylation and ubiquitination occur. Other chemical changes alter histones, such as histone lysine glucosamine acylation (GINA annylation), malonylation, butyrylation and crotonylation (ORNELLAS et al, 2017;CAVAGNARI,2012).

Depending on the chemical events (mentioned above) in the histone, there will be a lower affinity between the DNA and the histone being euchromatin, enabling gene expression;

Methylation and acetylation have opposite reactions, whereby an open and active chromatin with acetylated histones is interfered with by DNA-methyl-transferase, and Mecp2+ histone deacetylase ends up resulting in deactivated and condensed chromatin with closed nucleosomes and no possibility of gene expression, on the other hand, histone acetyltransferase causes histone demethylation, resulting in chromatin opening and activation, with open nucleosomes and the possibility of gene expression (ORNELLAS et al, 2017;CAVAGNARI,2012).

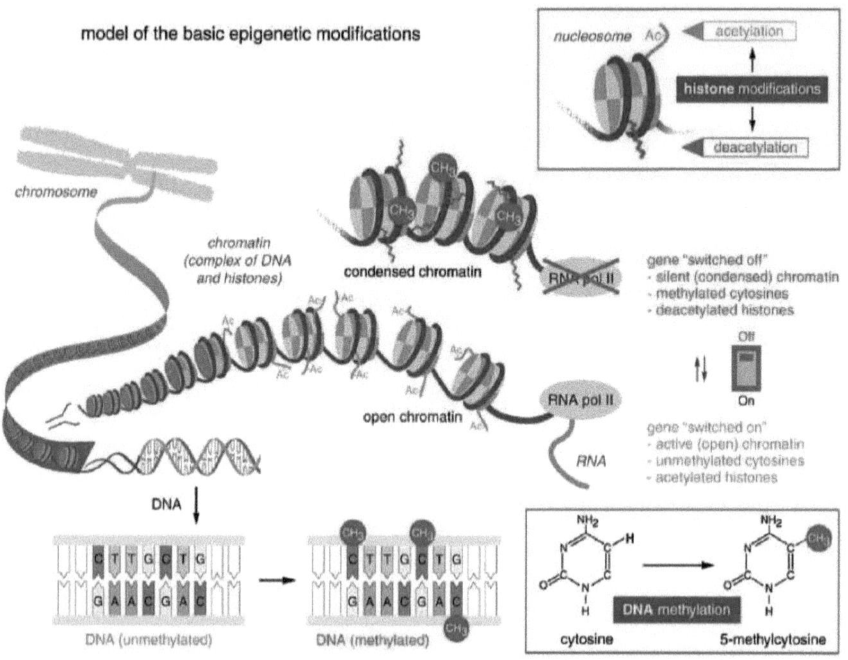

Figure 6 - Methylation - Acetylation

Source: VILCENSKAS; 2016.

CHAPTER 7

BIOTECHNOLOGICAL APPLICATIONS

In order to easily obtain stem cell sources with the potential of embryonic cells, stem cells have been developed with reverse induction through somatic cells; in addition, therapeutic cloning is being used where it is possible to obtain totipotent cells, with the possibility of having quantitative and qualitative cells (CHAPARRO et al, 2009; RADJI et al, 2009).

7.1 PLURIPOTENT STEM CELLS THROUGH REVERSE INDUCTION

Research carried out through the identification of Oct4, Sox2, c-Myc, and Klf4 factors, in which they develop a mechanism of dedifferentiation in which a somatic cell of a person, which may already be an adult, induces itself in reverse, transforming itself into a pluripotent cell, turning an already differentiated cell into a non-specialised pluripotent cell with immortal powers and the ability to replicate indefinitely, generating countless pluripotent cells (symmetrical division) or generating specialised pluripotent cells (asymmetrical division).

At the beginning of the research, induced pluripotency was obtained through a viral carrier, integrating the viral genome into the DNA of the recipient cells. Unfortunately, the splicing of the viral DNA led to the development of tumours and residual expression of exogenous factors.

Other methods have been developed, no longer using a viral vector, but using other substances as facilitators of entry into the target cell, through arginine tails (cell-penetrating peptides), which are overexpressed in the

HEK293 cell line, making it possible to induce pluripotency inversely through penetrating proteins together with transcription factors, inducing the activation of genes that promote pluripotency and deactivating differentiation genes (CHAPARRO et al., 2009; RADJI et al, 2009).

7.2 CLONING FOR THERAPEUTIC PURPOSES

Cloning occurs naturally in bacteria and consists of creating identical daughter cells. In synthetic methods, after several attempts, the first clone was made using material from a sheep, the famous Dolly case. The cloning experiment consists of two forms: reproductive cloning and therapeutic cloning (RADJI et al, 2009; ZATZ, 2004).

Reproductive cloning consists of discarding the nucleus of an egg and inserting in its place a nucleus from another somatic cell, which may even be from another individual. After this egg with the injected nucleus is introduced into a surrogate womb, it can generate an embryo and then a foetus that is born identical to the individual with the nucleus of the somatic cell (ZATS, 2004).

Therapeutic cloning, on the other hand, consists of removing the nucleus of a specific somatic cell from a patient who needs a certain regenerative treatment, after which it is placed in the place of the nucleus of the egg that was removed, and several divisions of these cells take place, assuming a totipotent behaviour, after which they are used for therapeutic purposes (VARELLA, 2004; ZATS, 2004).

There are several advantages to this method: there is no rejection, because the patient's own material was used; in quantitative terms, the necessary amount can be replicated; in qualitative terms, the totipotent cell has the same behaviour as the natural one, with good replication and differentiation, but it is not known about its *telomeres, taking* as an example the cloning of Dolly the sheep, whose *telomeres were used* up prematurely. As far as laws are concerned, this method is not always permitted, as it is sometimes associated with controversy, where it has already been banned in several countries (WALTERO et al, 2014; RADJI et al, 2012; RAMOS).

7.3 CLINICAL APPLICATIONS OF STEM CELLS IN TISSUE ENGINEERING

In order to produce tissues and organs in vitro, various types of polymeric biomaterials are being developed with the function of supporting and shaping an organ in which cells are cultured between these polymers, which are temporary, They are biodegradable to the extent that the tissue cells specialise and form the tissue. These polymers degrade into macromolecules, being a bioabsorbable material, eliminated by metabolic means without leaving collateral residues, or bioreabsorbable when their macromolecules are excreted, both of which have low incompatibility rates. They can also interact, causing a physiological response such as growth and differentiation at the implantation site (DOMINGUES et al, 2016; BARBANTI et al, 2005; SALOMÃO et al, 2011).

Among the most widely used bioresorbable polymers are a-hydroxy derivatives, poly(D-lactic acid), poly(DL-lactic acid), poly(glycolic acid) and polycaprolactone, which are degraded in a balanced way by hydrolysis (BARBANTI et al, 2005; SALOMÃO et al, 2011; CARDOSO et al, 2009).

An experiment was carried out to synthesise a skin tissue supported and structured by polymer frameworks, in which cytochemical stimuli were used to differentiate mesenchymal cells taken from the adipose tissue of a patient's abdomen, after which they were cultivated in two bi-laminar structures composed of PVA/collagen-AH at a concentration of 5x10.4 cells/ml; after 14 days the PVA membrane resulted in a smooth surface that mimicked the skin and was used as an analogue for the epidermis, while the collagen/AH framework was porous, allowing interaction between the cells. In terms of the mechanical characteristics of the tissue obtained, PVA resulted in elasticity of 2x10.5 N/m2, similar to that of human skin (DOMINGUES et al).

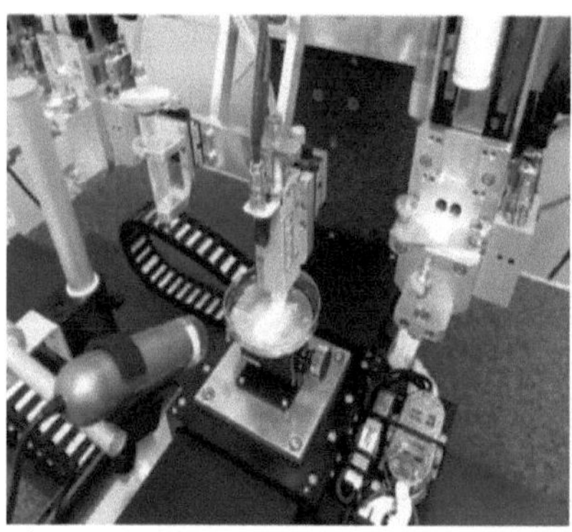

Figure 7 - Bone and ear printer and printout
Source: Ferrira, Nicolau.

7.4 CRISPR/CAS APPLICATION   IN STEM CELLS

The CRIS/PPR/CAS application is a technique where you can remove a defective gene, replace it by removing one and embedding another gene, or just insert a gene. The CRISPR enzyme and CAS9 originate from some bacteria where they have an immunological function in which CAS9 has a certain memory against the genetic code of bacteriophage viruses, identifying an invading viral DNA and degrading it (GONSALVES; et al; 2017).

CRIS/PR can be used in germ cells to correct certain genetic errors that will be inherited from one of the parents and prevent this disease from manifesting itself in future generations; it can also be used to correct somatic cells by correcting or exchanging a gene, but it is not possible to pass it on to future generations (GONSALVES; et al; 2017).

CHAPTER 8

STEM CELL SOURCES

Chart 3 - Stem Cells

Source of stem cells:	Type of stem cell				Cell bank
	Powerful Toti	Pluri potent	Multi powerful		
			Haematopoietic	Mesenchymal	
Milk pulp		X		X	X
Liqui/ amniotic				X	
Umbilical cord			X		X
Peripheral blood			X	X	X
Embryo		X			X
Adipose				X	
Therapeutic cloning	X	X			
Reverse induction		X			

CHAPTER 9

SOME REGENERATIVE TREATMENTS WITH STEM CELLS

There are several regenerative stem cell treatments that have already achieved a certain satisfactory goal; some using stem cell banks, others using the patient's own multipotent stem cells, but in both cases the correct induction is a challenge. The improvement of this treatment is hopeful for the development of control and cure of diseases such as Alzheimer's and Parkinson's.

9.1 REGENERATIVE APPLICATIONS IN CARDIOVASCULAR TISSUES

In a case report, through an experimental treatment carried out at the Pró-cardíaco and Miguel Couto hospitals, 24 patients with heart attacks underwent removal of cells from their own bone marrow; afterwards, a guide catheter was placed up to the coronary artery related to the heart attack, where stem cells were injected over a certain period of time, appearing to be a viable and safe result with good regeneration of the damaged tissue, but there is a need for more studies on the application (ALVES et al, 2017).

Only two sources of stem cells have been used in patients so far, those from skeletal muscle (origin of myoblasts) and bone marrow, promoting great chances of therapies, but little is known about the best route for transplantation, injected concentration, whether certain treatment and culture of these cells with supposed activators in the desirable tissue

(CARVALHO et al, 2009; MOEIRA et al, 2011).

9.2 REGENERATIVE APPLICATIONS IN NEUROLOGICAL TISSUES

In an experiment carried out on Wistar rats subjected to spinal cord injury, some were given no treatment at all and others were treated with human umbilical cord stem cells, with concentrations of haematopoietic and mesenchymal stem cells. In some rats, five days after the spinal injury, the cells were injected intravenously, improving behavioural deficits, and the differentiation of some umbilical cord cells into neuronal cells was observed, improving axonal regeneration and motor function.

On the other hand, it was noted that rats that were injected with stem cells immediately after the injury had lower results than those that were injected with stem cells after nine days. In hypothesis, this result after nine days was obtained due to inflammatory processes that favour the survival of the transplanted stem cells. In comparison, the rats that received treatment all had significant functional recovery far above those that did not receive stem cell treatment (RODRIGUES et al, 2012).

Other neurological diseases include Alzheimer's and Parkinson's, both of which are incurable and both of which involve the degeneration of neuronal tissue, with stem cell therapy being the best hope of treatment. Parkinson's disease is characterised by trembling and poor motor function. These symptoms are caused by the apoptosis of nigra neurons, which are responsible for the production of dopamine, an essential substance in the synapses of movement.In Alzheimer's, an accumulation of b-meloid proteins occurs irregularly, forming plaques and inflation, causing neuronal degeneration and gradual memory loss (BETTENCOURT et al, 2008;

PEREIRA, 2008; MUOTRI, 2017; SENNEN et al, 2011).

9.3   DIABETES MELLITUS 2

It can act on two pathogenic mechanisms: resistance to the action of insulin and secretory dysfunction of b-pancreatic cells, occurring either genetically or due to environmental factors such as obesity. Through oxidative stress, amyloid deposition, glucotoxicity, lipotoxicity, inflammation and stress on the rough endoplasmic reticulum, pancreatic b-cells undergo apoptosis, reducing their mass and insulin secretion (GIMENO et al, 2011).

Through the use of pruripotent and multipotent stem cells, research has been carried out with the aim of increasing the quantity and functionality of pancreatic b-cells in rats with diabetes mellitus 2. Soria et al. demonstrated **that** pruripotent cells implanted in the spleen of rats differentiated into insulin-secreting cells and were capable of producing normoglycaemia (GIMENO et al, 2011).

Kojima et al, observation of extra pancreatic insulin-producing cells in hyperglycaemic rats treated with bone marrow-derived multipotent stem cells.

Ende and colleagues achieved a reduction in glycaemia, improved survival and attenuation of kidney damage in obese rats through treatment with stem cells derived from the human umbilical cord. Good results have been obtained in various experiments on rats, but few cases have been reported in humans, perhaps because it is not feasible due to difficulties such as synchronising the cells induced to produce insulin with the need for metabolism, and there is currently little knowledge about the mechanism of differentiation, proliferation, regeneration and plasticity of pancreatic b-cells, nor is it known whether there is a pancreatic b-cell progenitor stem cell or whether they are self-renewing (CRISTANTE et al, 2011; LEAL et al, 2010; GIMENO et al, 2011).

CHAPTER 10

CONTROVERSIES AND LAWS REGARDING STEM CELL MANIPULATION

Due to religious sentiment and supposedly ethical thoughts, perhaps due to a lack of information, there are times when stem cell research is controversial in relation to its manipulation; because it is the most viable for research and therapeutic use, it is necessary to destroy the embryo in order to obtain it, which is why the right to life of the supposed embryo is being discussed, considering it a human being.

According to Law No. 10.406 of 10 January 2002art. 2[1] , "the civil personality of the person begins after birth with life; but the law protects, from conception, the rights of the unborn."

From the moment of birth, rights and duties are acquired, but the law also protects the rights of the unborn child, who is considered to be a human being whose birth is expected to be certain and who is developing in a uterus. Various biosafety laws have been developed with advances in science, such as in vitro fertilisation, in which embryos are formed in the laboratory in order to select the best ones for a possible pregnancy in the patient, those that remain are frozen and discarded after three years; in the case of this manipulation of in vitro fertilisation, the law states that the resulting embryo without having been transferred to the mother's womb is neither a person nor an unborn child. The aim of research and clinical applications is to use non-viable embryos that have a risk of miscarriage or chromosomal malformation and to use embryos frozen after three years, all

---

[1] Civil Code - Law 10406/02. Available at < https://www.jusbrasil.com.br/topicos/10731286/lei-n- 10406-de-10-de-janeiro-de-2002 > Accessed on 30.nov.2017.

donated by the parents (12,59).

In Brazil, even after much debate and attempts to ban the use of embryonic cells, Brazilian law[2] imposes that:

> Law 11.105, enacted on 24 March 2005, on biosafety, states in article 5 that the use of embryonic stem cells obtained from human embryos produced by in vitro fertilisation and not used in the respective procedure is permitted for research and therapy purposes.

The embryos must be non-viable or frozen for more than three years, and the embryo's parents must also donate them (12,58). The bioethics launched by Van Rensselaer Potter believes that the surplus embryos allowed by the biosafety law, and the release of embryos for therapy and research, must be done by the parents, who are responsible for their moral and religious convictions (12). In international legislation, the United States and the United Kingdom allow embryonic stem cell research and therapeutic cloning. France, Holland, Spain, Portugal, Switzerland and Australia allow embryonic stem cell therapy, but do not allow therapeutic cloning.

In Italy, due to pressure from the Catholic Church, embryonic stem cell research is not allowed, while in Germany it is only permitted with imported stem cells. Human cloning is banned in all countries (12).

In Brazil, Law 11.105 of 24 March 2005 establishes that the use of discarded and frozen embryos is permitted for research and therapeutic purposes after three years without any intention of use, also with the proper release of the parents; therapeutic cloning is also permitted by the same law,

---

[2] BRAZIL, Law n.11.105. Available at <http://www.planalto.gov.br/ccivil 03/ ato2004-2006/2005/lei/l11105.htm> Accessed on: 30.nov.2017

with reproductive cloning being prohibited (48,60,61,62).

CHAPTER 11

METHODOLOGY

Bibliographical research of scientific articles in portals and scientific journals.

CONCLUSION

Among various aspects, it is necessary to better understand the type of stem cell and its application: a) embryonic stem cells obtain more effective results with fewer risks, are easier to manipulate and can be used over a long period of time and with numerous replications; b) multipotent stem cells, unlike embryonic stem cells, are more limited and less controlled, subdivided into mesenchymal stem cells, which are difficult to control over a long period of time and may have undesired differentiation, and haematopoietic stem cells, which are more limited to the formation of blood cells and the immune system and have their telomeres already reduced.

The use of a) cells is more effective and easier to manipulate, but it is not always possible to use them due to the need to destroy the discarded embryos. This leaves two other ways of obtaining pluripotent cells, therapeutic cloning, which also ends up being involved in controversy in the same way as embryos, leaving only the obtaining of pluripotent cells through the reverse induction of somatic cells, in both of which rejection is negligible because it is the patient's own material, but more in-depth studies are still needed.

On the other hand, b) cells are easier to access and have several sources in an adult organism, can be removed several times and there is the possibility of cell banks. There have been several reports of success in various types of manipulation of these cells, but it is worth mentioning the risks that can occur, such as chimerism, tumour cells, cellular senescence, cellular apoptosis, shortening of telomeres with lower than expected results or side-effects, so there is a need for greater mastery of these cells through genetic preservation via the telomerase enzyme, an understanding of epigenetics and its interaction with substances that would help with gene control.

In Brazil, the use of stem cells from frozen embryos that have been discarded and donated by the parents and not allowed to be commercialised is allowed. Therapeutic cloning is also allowed, but without the intention of implantation in the uterus.

REFERENCES

ABDELHAY, et al Stem cells of haematopoietic origin: expansion and prospects for therapeutic use. **Brazilian Journal of Haematology and Haemotherapy**. V.31 supl.1 São Paulo, 2009.

AFANADOR, Carlos Humberto; PENA, Carlos Mario Munetón. Epigenetics of colorectal cancer. **RevColgastroenterol.** v.33 n.1 Bogotá, jan./mar, 2018.

ALVES, Endrigo P. et al. Isolation and cultivation of mesenchymal stem cells extracted from adipose tissue and bone marrow of dogs. **Ciência Animal Brasileira**. V.18. Goiânia, 2017.

ARAGÃO, Maria A. Do Carmo; Bezerra, Francisco T. Gomes;. Brazil and stem cell research: an overview. **Journal of biology**. v.34. São Paulo: 2012.

BARBANTI, et al.Bioreabsorbable Polymers in Tissue Engineering.

Journal of Science and Technology. v.15. n.1. 2005, p.13-21,

BARRETO, Madson A. et al. The consequences of the decrease in dopamine produced in the substantia nigra: a brief reflection. **Interfaces Científicas-Saúde e ambiente**. Aracatu. v.4. n.1. p.83-90. Oct.2015.

BARROSO, Luís Roberto; Gestation of Anencephalic Foetuses and Stem Cell Research: Two Themes on Life and the Dignity of the Constitution. **Psnóptica**. Year 1, n. 7, April 2007.

BENÍTEZ, A; et al. Endocrine response to the application of whole-body vibration in humans. **Med deporte**. v.8 n.3 Sevilla, 2015.

BETTENCOURT, Rosane; fogliato, Laura; Et al. Haematopoietic stem cell transplantation in Hodgkin's lymphoma.**Revista Brasileira de Hematologia e Hemoterapia**. v.31 supl.1 São Paulo: 2009.

BYDLOWSKI, Sergio P.; debes, Adriana A.; Et al. Stem cells from amniotic fluid. **Revista Brasileira de Hematologia e Hemoterapia**. v.31 supl.1 São Paulo,2009.

BRAZIL. Federal Law No. 11.105, of 24 March 2005. Establishes safety standards and inspection mechanisms for activities involving genetically modified organisms. **Presidency of the Republic**. 24 March 2005.

CAMPOS, Lorena L.; Alvarenga, Fernanda L.C; Et al. Isolation, culture, characterisation and cryopreservation of stem cells derived from mesenchymal amniotic layer and umbilical cord tissue of bovine fetuses. **Pesq. vet.bras.** v.37 n. 3 Rio de Janeiro: 2017.

CAR, Steven M. Flurescent In-situ Hybridisation (fish) identification of Human chromosomes-"chromosome Painting". **Text & image.** 2008.

CARDOSO, Guinea B. Camargo; Arruda, Antonio C. Fonceca. The role of stem cells in tissue engineering. **Ciências e Cognição**. v.14. n.3 Rio de Janeiro: 2009.

CARVALHO, Antonio C.C.C; et al.Bases of cell therapy in cardiovascular. **Ver. Brás .Hematol. Hemoter.** v.31 supl.1 São Paulo, 2009.

CAVAGNARI, Brian. M; Regulation of gene expression: how epigenetic mechanisms work. **Archivos argentinos de pediatria**. Arch. Argent. Pediatr.v.110 n.2 Buenos Aires: 2012.

CHAPARRO, Orlando; Beltrán, Orietta;. Nuclear reprogramming and induced pluripotent cells. **Med Magazine**.v.17 n.2.Bogotá, July/December, 2009.

COLPO, et al. Mesenchymal stem cells for the treatment of neurodegenerative and psychiatric diseases. **Canais da Academia Brasileira de Ciências**. v.87 n.2 supl.0 Rio de Janeiro, 2015.

CRISTANTE, Alexandre Fogaça; Narazaki, Douglas Kenji. Advances in the use of stem cells in orthopaedics. **Ver. Brás. Ortop**. V.46. n.4. São Paulo, 2011.

DOMINGUES, J.A; Hausen.M; Et al.Hybrid Pva/Collagen/Hyaluronic Acid Device Cultivated with Mesenchymal Stem Cells for Dermal Regeneration. **Biomaterials Laboratory, Faculty of Medical and Health Sciences**.PUC/SP.

ELORRIAGA, Tomas; Epigenetic mechanisms. **Humaning ES blog**. 2018/05/03.mecanismos-epigenéticos.

FONTENELE, Eveline G. et al. Environmental contaminants and endocrine interferents. **Arquivos Brasileiros de Endocrinologia & Metabologia**. v.54 n.1 São Paulo, 2010.

GAMBOA, Karla A Bascunan; QUEZADA, Magdalena Araya; ET AL;.MicroRNAs, epigenetic mechanism to study coeliac disease. **Revista espanola de enfermedades digestivas**. v.106.n.5 Madrid: 2014.

GONSALVES, Giulliana A. Raquel; PAIVA, Raquel M. Alves;.Gene therapy: advances, challengesand perspectives. **Reviewingbasic sciences**. Eisntein. Vol.15 no.3 São Paulo july/sept.2017.

GÓMEZ, Diego L. et al. Telomerase and telomere: their structure and dynamics in health and disease. **Medicina (B.Aires)**. v.74 n.1 BuenosAiresJan. /Feb.2014.

GOMES, Kátia M. et al. Induced pluripotent stem cells: the role of epigenetics in reprogramming and its clinical applicability. **Journal of the Brazilian Medical Association**. v.63 n.2. São Paulo: 2017.

JUNIOR, Francisco C. da Silva; et al. Haematopoietic stem cells: uses and perspectives. **Revista Brasileira de Hematologia e Hemoterapia**. v.31 supl.1 São Paulo: 2009.

KABAYASHI, Natália C. Cristina; NORONHA, Samuel M. Ribeiro.Cancer stem cells: a new approach to tumour development. **Revista da Associação Médica Brasileira**.v.61. n.1 São Paulo: 2015.

LEAL, AngelaM.O; Voltarelli, Júlio César;.Perspectives on stem cell therapy for type 2 diabetes mellitus. **Rev. Bras. Hematol**. v.32. n.4 São Paulo: 2010.

MAGALHÃES, Georgia M; et al;. Evaluation of tumour stem cell immunolabelling in mammary carcinosarcomas and carcinomas in mixed tumours in bitches. **Pesquisa Veterinária Brasileira**. v.34. n.5. Rio de Janeiro: 2014.

MAZZARELLA, Rosane; GARNICA, Taismara K.; et al. Stem cells derived from the olfactory epithelium: therapeutic perspectives in veterinary medicine. **Pesq. Vet. Bras.** v.36. n.8. Rio de Janeiro: 2016.

MIRANDA, Maribel Mata; Zapién, Gustavo J. Vazquez; Et al. Generalities and applications of mother cells. **Perinatol. Reprod. Hum**. V.27 n.3 México: 2013.

MONTOYA; GINER; et al. What are microRNA?Possible biomarkers and therapeutic targets in osteoporotic disease. **Bone metabolism unit**. 2013

MOREIRA, Rodrigo C; Haddad, Andréa F.; Et al;.Intracoronary injection of stem cells after myocardial infarction. **Microcirculation substudy**. v.97. n.5 São Paulo: 2011.

MURANSKI, Pawel; RESTIFO, Nicholas P;.Essentials of Th17 cell commitment and plasticity. **Blood Advertisement**.

ORNELLAS, Fernanda; CARAPETO, Priscila; ET AL;.Obese fathers lead to na altered metabolism and obsity in their children adulthood: review of experimental and human studies. **Journal of paediatrics. J.pediatr.**v.93. n.6 Porto Alegre: 2017.

PAYÃO, Spencer L.M; SEGATO, Rosimeire; Et al.Genetic control of cultured human stem cells. **Ver. Brás. Hematol. Hemoter**. V.31 supl.1 São Paulo, 2009.

PEREIRA,Lygia da Veiga. The importance of using stem cells for public health. **Ciênc. Saúde coletiva**. v.13 n.1 Rio de Janeiro: 2008.

PERINI, Silvana; Silla, Lúcia M.R.; Et al. Telomerase in haematopoietic stem cells. **Revista Brasileira de Hematologia e Hemoterapia**. v.30. n.1. São José do Rio Preto:2008.

PINHO, Mauro de Sousa Leite;.Tumour stem cells: a new concept in colorectal carcinogenesis. Brazilian Journal of Coloproctology. **Ver Brás. Colo-proctol**. v.29. n.1 Rio de Janeiro: 2009.

RADJI, Mohammed A. Mostajo; Ferreira, Leonardo. M.R;CambiandolaIdentidad celular para crear una verdadera

personalised medicine.**GacMed**
Bol. v.35 n.2 Cochabamba dic. 2012.

RODRIGUES, L.P; et al. Human umbilical cord blood mononuclears promote functional recovery after traumatic spinal cord injury in Wistar rats. **Braz. J MedBiol Res**. v.45. January, 2012.

SANDOVAL, Cristian; VÁSQUEZ, Bélgica; et al;. Role of alcohol consumption and antioxidants on global DNA methylation and cancer. **International journal of Morphology**. Int. J. Morphol. V.36 n.1 Temuco mar.2018

SANMARTÍN, Sánchez L.; et al. Evaluation of telomere length in horses using quantitative polymerase chain reaction. **Military health**.v.70. n.2 Madrid: 2014.

SEBBEN, Alessandra Deise; Lichtenfels, Martina; Et al. Peripheral nerve regeneration: cell therapy and neurotrophic factors. **Brazilian Journal of Orthopaedics**. Ver. Brás. Ortop. V.46 n.6 São Paulo, 2011.

SILLA, Lucia M. R.; DULLEY, Frederico; ET AL. Haematopoietic stem cell transplantation and acute leukaemia: Brazilian guidelines. **Revista Brasileira de Hematologia e Hemoterapia**.v.32 supl.1 São Paulo, 2010.

SOLOMÃO, Zenaide; Dias, Carmem, G.B.T.; et al.Evaluation of **a pcl and 0-tcp biotissue.Proceedings of the 11th Brazilian Polymer Congress**. Campos do Jordão, SP - 16 to 20 October 2011.

SOUZA, Cristiano Freitas; et al. Mesenchymal stem cells: ideal cells for cardiac regeneration? **Rev. Bras. Cadiol**. Invasiva v.18 n.3 São Paulo, 2010.

TORRÃOL, Andréa et al.**Different approaches, one target**: understanding the cellular mechanisms of Parkinson's and Alzheimer's diseases. University of São Paulo. USP.

VARELLA, Dráuzio. Human cloning. **Estudos Avançados**. v.18 n.51 São Paulo, 2004.

VIGORITO, Afonso C.; Souza, Carmino A. Haematopoietic stem cell transplantation and the regeneration of haematopoiesis. **Revista Brasileira de Hematologia e Hemoterapia**.V.31 supl.1 São Paulo: 2009.

VILLAMARÍN, Diego Germán Pérez. Evolution, embryonic development and psychism. **Revista Latinoamericana de Bioética**. v.9 n.2 Bogotá: 2009.

VISENTAINER, Jeane E. Laguila; Sell, Ana Maria; et al. Importance of cytokine regulatory gene polymorphisms in haematopoietic progenitor cell transplantation. **Revista Brasileira de Ciências Farmacêuticas**. v.44. n. 4 São Paulo: 2008.

WALTERO, Edgar M. Mongollón; Mello, Marco R. Bourg; Et al. Cloning bovine embryos from somatic cells. **Orinoquia**. v.18. n.1 Meta Jan./June 2014.

YARAK, Samira; Okamoto, Oswando, Keith;. Stem cells derived from human adipose tissue: current challenges and clinical perspectives. In. **Bras. Dermatol**. V.85 n.5 Rio de Janeiro Sept./Oct.2010.

ZATZ, Mayara;. Cloning and stem cells. **Advanced Studies**. Estud. Av. v.18 n.51 São Paulo: 2004.

FAUSTO, Caemen C. do Val; CHAMMAS, Maria Cristina; ET AL;. Thymus: ultrasound characterisation. **Radion Brás.** 2004; 37(3) 207-210.

ZANOTO, Alfeu Filho;. Differential effects of retinol and retinoic acid on cell proliferation, death and differentiation: the role of mitochondria and xanthine oxidase in the pro-oxidant effects of vitamin A. **LUME** . 2009

NUNES, Maria Tereza;. Thyroid hormones: mechanism of action and biological importance. **Arquivos brasileiros de endocrinologia & metabologia.** Arq Brás endocrinol metab. Vol.47 no.6 São Paulo dec.2003

DUSSO, Adriana S; BROWN, Alex J.; ET AL;.Vitamin D. **America journal of physiology.** 336. 2004.

MACIEL, Lea M. Zanine; MAGALHAENS, Patricia K. R.;.Thyroid and pregnancy.**Arquivo brasileiro de endocrinologia e metabologia.** Arq Brás endocrinoç metab.vol 52 no.7 São Paulo or. 2008.

VILCINSKAS, Andreas;. The role of epigenics in host-parasite coevolution: lessons from the model host insects galleria mollonela and tribolium castaneum. **Zoology**. 19 (2016) 273280.

More
Books!

OMNIScriptum

Printed by Books on Demand GmbH, Norderstedt / Germany